COMPETITIVE CHEMISTRY

INTRODUCTION

This objective chemistry series provides a basic and challenging problem of chemistry from particular topics. It can be used to brush up ones basics and checking up the preparation level of particular topic. It is equally helpful to the traditional classes as well as competitions. It can be also taken as a revision material for any competition which includes the test of basic chemistry. If you want to grasp the subject before practicing these multiple choice questions, you can go through the website http://www.ncert.nic.in/ncerts/textbook/textbook.htm and down load the free copy of science books and after having command on the topic practice it.

If you have any query or suggestion about this series you can send your suggestion at uk2594@gmail.com.

CONTENTS

CHAPTER 1   MATTER IN OUR SURROUNDINGS

CHAPTER 2   IS MATTER AROUND US PURE

CHAPTER 3   ATOMS AND MOLECULES

CHAPTER 4   STRUCTURE OF ATOM

CHAPTER 5   CHEMICAL REACTION AND EQUATIONS

CHAPTER 6   ACID BASES AND SALTS

CHAPTER 7   METALS AND NON METALS

CHAPTER 8   CARBON AND ITS COMPOUNDS

CHAPTER 9   PERIODIC CLASSIFICATION OF ELEMENTS

1. MATTER IN OUR SURROUNDINGS

SOME IMPORTANT POINTS

➢ The things that occupy space and have mass are known as matter.

➢ A matter is made up of very small particles.

➢ These p[articles of matter have spaces between them, these are continuously moving, these attract each other.

➢ The intermixing of these particles in gaseous phase is known as diffusion.

➢ Simply matter is found in three states: solid, liquid and gases.

➢ The force of attraction is maximum in solids and least in gases.

➢ The intermolecular space is maximum in gases and least in solids.

➢ The states of matter can be changed by changing the temperature or pressure.

➢ In sublimation solid directly changes into gases.

➢ The phenomenon of changing of a liquid into its vapour at any temperature below its boiling point is known as evaporation.

➢ The rate of evaporation depends upon the surface area, temperature, humidity and the speed of wind.

➢ Evaporation causes cooling.

1. MATTER IN OUR SURROUNDINGS

1. Everything in this universe is made up of?

    a) Metal   b) Non metal   c) Matter   d) None of these

2. What is the example of matter?

    a) Star   b) Plants   c) Clouds   d) All of these

3. Five basic elements are?

    a) air, Earth, fire, sky and water   b) plant, air, sky, soil &Matter

    c) fire, material, earth, land   d) None of these

4. How Many types of classification of matter are?

    a) Two   b) three   c) four   d) five

5. Matter is made up of

    a) soil   b) rock   c) Particles   d) none of these

6. What is the colour of water?

    a) Red   b) Sky colour   c) colourless   d) golden

7. The particles of matter are?

    a) small   b) long   c) very small   d) None of these

8. Space between particles of one type of matter are?

    a) big space b) small space   c) enough space d) not space

9. Particles of Matter are -------------- moving.

    a) Straight line   b) continuously   c) both   d) None of these

10. As the temperature rises, particles move?

     a) slowly-slowly   b) faster   c) None of these

11. The kinetic energy of the particles ………………. On increasing temperature.

a) decreases   b) move backward c) increases d) None of these

12.   Intermixing of particles of two different type matters on their own is called?

a) Collision   b) Matter   c) Particles   d) diffusion

13.   Particles of matter attract:

a) each other   b) together   c) other object   d) None of these

14.   Which states of matter arise due to the variation in the characteristics of the particles of matter?

a) solid   b) Liquid   c) gas   d) all of these

15.   What are the example of solid, liquid, gas:

a) rock, water oxygen   b) soil, tea, coffee

c) H_2, H_2SO_4, Nitrogen   d) None of these

16.   Solids have a tendency to maintain their:

a) Shape   b) Volume   c) Velocity   d) Both a and b

17.   Liquids have no fixed shape but have a:

a) fixe volume   b) Fixed density   c) None of these

18.   Liquids flow and change shape so they are not rigid but can be called?

a) diffusion   b) fluid   c) matter   d) None of these

19.   Solids and liquids can diffuse into:

a) solids   b) liquid   c) gaseous   d) None of these

20.   the gases from the atmosphere diffuse and dissolve in?

a) Water   b) salt   c) snow   d) None of these

21.   Solids, liquids and gases can diffuse into?

a) liquids   b) solids   c) gases   d) all of these

22. The rate of diffusion of liquid is higher than that of?

    a) gases   b) solids   c) none of these   d) a & b

23. Gases are highly compressible as compared to?

    a) solids and liquids   b) solids and gases

    c) liquies and gases   d) None of these

24. CNG is used as?

    a) Fuel   b) Breathe   c) Both a and b   d) None of these

25. The aquatic animals can breathe under water due to the presence of dissolve oxygen in?

    a) Earth   b) under the Earth   c) Water   d) None of these

26. Define density?

    a) The mass per unit volume of a substance

    b) The mass per unit volume square of a substance

    c) Both a and b

    d) None of these

27. On increasing the temperature of solids which energy of the particles increases

    a) Kinetic energy   b) Potential energy

    c) Mechnical energy   d) None of these

28. Due to the increase in kinetic energy the particles stars:

    a) Moving   b) vibrating   c) None of these

29. The energy supplied by heat overcomes the force of attraction between the

    a) Matter   b) object   c) particles   d) all of these

30. The temperature at which a solid melts to become a liquid at the atmospheric pressure is called its?

a) individual point b) point   c) centre   d) melting point

31.   The melting point of ice is?

      a) 273.16Kg   b) 273.16J c) 273.16N   d) 273.16K

32.   The Process of melting, which is change of solid state into liquid state is also known as?

      a) fussion   b) fusion   c) collision   d) None of these

33.   When we supply heat energy to water particles, particles start moving even:

      a) faster   b) slowly   c) None of these

34.   The temperature at which a liquid starts boiling at the atmospheric pressure is called?

      a) boiling point   b) melting Point   c) latent heat
      d) None of these

35.   Boilling is as Phenomenon of?

      a) Albert   b) Bulk   c) microscopic   d) none of these

36.   Particles from the bulk of the liquid gain enough energy to change into the?

      a) Liquid state   b) solid state   c) gaseous state   d) vapour state

37.   The state of matter can be changed into another state by changing the?

      a) volume   b) Mass   c) temperature   d) None of these

38.   What is the SI unit of temperature?

      a) K   b) J   c) Kwh   d) None of these

39.   The SI unite of volume is?

      a) m^2 b) m^3 c) m/s   d) metre

40.   What is the common unit of volume is

      a) m^3 b) litre   c) metre   d) m/s

41. What is unit of pressure?

   a) pascal   b) Kwh   c) Kgm/s^2 d) None of these

42. The Pressure of air in atmosphere is called?

   a) atmospheric Pressure b) Normal Pressure

   c) Pressure   d) None of these

43. A change of state directly from solid to gas without changing into liquid is called?

   a) sublimation   b) vaporization   c) vibration d) None of these

44. The solid carbon dioxide is also known as?

   a) dry ice   b) fusion   c) None of these

45. The change of a liquid into vapors at any temperature below its boiling point is called?

   a) Evaporation   b) fusion   c) collision   d) None of these

46. The state of matter can be changed by changing temperature or?

   a) Volume   b) pressure   c) velocity   d) None of these

47. What is the unit of weight?

   a) K   b) Newton   c) Kg   d) Pa

48. What is the unit of length?

   a) meter   b) meter/second   c) m/s^2 d) None of these

49. The particles gain energy from your palm or surrounds and evaporate causing the palm to feel?

   a) Cool   b) hot   c) both   d) None of these

50. During summer we perspire more because of the mechanism of our body which keeps us?

   a) Cool   b) hot   c) None of these

51. The sun and the stars glow because of the presence of:

    a) Plasma   b) heat   c) light   d) None of these

52. Who had done some calculations for a fifth state of matter?

    a) S.N. Bose   b) Isaac Newton   c) None of these

53. When Indian physicist S.N. Bose had done some calculations for a fifth state of matter

    a) 1930   b) 1920   c) 1940   d) 1950

Answers:

Ques.	Ans.	Ques.		Ques.		Ques.		Ques.		Ques.	
1	C	11	C	21	A	31	D	41	A	51	A
2	D	12	D	22	B	32	B	42	A	52	A
3	A	13	A	23	B	33	A	43	A	53	B
4	A	14	D	24	A	34	A	44	A		
5	C	15	A	25	C	35	B	45	A		
6	C	16	D	26	A	36	D	46	B		
7	C	17	A	27	A	37	C	47	B		
8	C	18	B	28	B	38	A	48	A		
9	B	19	B	29	C	39	B	49	A		
10	B	20	A	30	D	40	B	50	A		

2. IS MATTER AROUND US PURE

SOME IMPORTANT POINTS

- Mixture is a combination of two or more substances in any proportion.
- Mixtures components can be separated using appropriate technique.
- In homogeneous mixture each and every part of mixture shows uniform composition while it is non uniform in non homogeneous.
- Solution is a homogeneous mixture. The major components of solution are known as solvent while minor is termed as solute.
- Colloids and suspensions are heterogeneous mixture of substances. In suspension the solute particles are seen by naked eye. In colloids the solute particles are not seen by naked eye but can scatter the light.
- We can separate the dye from ink evaporation.
- Configuration is useful for separating cream from milk.
- Two immiscible liquids are separated using separating funnel.
- Mixture of Ammonium and salt can be separated by sublimation.
- Mixture of two miscible liquids can be separated by fractional distillation method.
- Air is a homo9geneous mixture of gases and these components can be separated from air.
- Crystallization technique is useful to obtain pure copper sulphate from impure sample.
- During the burning of candle both chemical and physical changes occur.
- Elements and compounds are the pure substances. Properties ,of compounds are different from its constituents.

2. IS MATTER AROUND US PURE

1. What is a mixture?

   a. Mixtures is constituted by one kind of pure matter.

   b. Mixture is constituted by more than one kind of pure or impure matter.

   c. Both a and b

   d. None of these

2. From which process dissolved sodium chloride can be separated from water?

   a. Centrifugation   b. Filtration   c. Evaporation   d. Distillation

3. Sugar is a?

   a. Pure matter   b. Impure Matter   c. Mixure   d. None of these

4. What is a solution?

   a. a heterogeneous mixture of two or more substances.

   b. a pure form of matter

   c. Homogeneous mixture of two or more substances

   d. None of these

5. Which of the following is a solution?

   a. sugar   b. Kerosene oil   c. water   d. Soda water

6. Which of the following is not a solution?

   a. Soda water   b. Lemonade   c. Air   d. None of these

7. What is alloy?

   a. mixture of two or more metals or a metal and non metal

b. mixture of two or more metals

c. Mixure of two or more non metals

d. None of these

8. Which cannot be separated by physical method?

a. Solder b. Bronze c. Brass   d. All of these

9. Brass has the % of zinc and copper respectively:

a. 70, 30 b. 50, 50 c. 30, 70 d. None of these

10. In a solution the component present in larger amount is known as?

a. Solute   b. Solvent   c. Solution   d. All of these

11. In a Solution the compound present in lesser quantity is known as?

a. Solute   b. Solvent   c. Solution   d. All of these

12. A solution of iodine in alcohol is known as ?

a. Compound iodine   b. Tincture of iodine

c. Pure form of iodine   d. None of these

13. Air is a homogenous mixture of a number of gases .its two main constituents are ?

a. Oxygen (21%) and Nitrogen (78%)   b. Oxygen (20%) and Nitrogen (78%)

c. Oxygen (22%) and Nitrogen (78%)   d. Oxygen (21%) and Nitrogen (77%)

14. A solution is _____ mixture .

a. Heterogeneous mixture   b.Colloidal mixture

c. Suspension mixture   d. Homogeneous mixture

15. 1nm =?

a. 10^{-7}m   b. 10^{-9}m   c. 10^{-8} m   d. 10^{-10}m

16. Which can scatter beam of light?

    a. Colloidal b. Solution c. Suspension   d. All of these

17. In which solute particle settle down when left undisturbed?

    a. Solution b. Suspension c. Colloidal   d. Both (b) &(c)

18. The amount of the solute present in the saturated solution at particular temperature called its _____ .

    a. Solubility b. Solvity   c. Colloidility   d. None of these

19. Concentrated of solution = ?

    a. Amount of solute /Amount of solution   b. Amount of solute/ Amount of solvent

    c. Both (a)&(b)   d. Amount of solvent /Amount of solute

20. Mass by mass percentage of a solution =?

    a. Mass of solution /Mass of solute *100   b. Mass of solution /Mass of solute *1000

    c. Mass of solute /Mass of solution*1000   d. Mass of solute /Mass of solution*100

21. Mass by volume percentage of solution =?

    a. Mass of solution /Volume of solution*100

    b. Mass of solution/volume of solute*100

    c.Mass of solute /volume of solution*100

    d. None of these

22. A solution contains 30g of sugar in 430 g of water .Calculate the concentration in terms of mass by mass percentage of the solution ?

    a. 6.78%   b. 6.52%   c. 7.52%   d. None of these

23. In which of the following we can see the particle of substance to the naked eye?

a. Suspension   b. Colloidal   c. Solution   d. None of these

24. Which of following is a heterogeneous?

a. Suspension   b. Colloidal   c. Both (a)&(b)   d.None of these

25. The scattering of a beam of light is called _____.

a. Tincture effect   b. Tyndall effect   c. Scattering   d. Natural effect

26. From which technique we can separate the colloidal particle ?

a. Sublimation   b. Evaporartion   c.Filtration   d. Configuration

27. In which of the following is a colloidal ?

a. Cheese   b. Chips   c. Ammonium chloride   d. None of these

28. In which of the following is a type of colloidal?

a. Aerosol   b. Emulsion   c. Sol   d. Gel

29. When dispersed phase is solid and dispersing medium is liquid the type of colloidal is?

a. Solid sol   b. Emulsion   c. Sol   d. Foam

30. Shaving cream is example which type of colloids?

a.Aerosol   b. Sol   c. Solid Sol   d. Foam

31. How can we obtain coloured component (DYE) from Blue /Black ink ?

a. Evaporation   b.Sublimation   c. Both (a)&(b)   d. Filtration

32. How can we separate cream from milk?

a. Evaporation   b. Centrifugation   c. Separating funnel   d. None of these

33. Which of the following method used in diagnostic laboratories for blood and urine test ?

a. Centrifugation   b. Evaporation   c. Filtration   d. All of these

34. How can be we separate a mixture of two immiscible liquids ?

   a. Evaporation   b. Sublimation   c. Filtration   d. None of these

35. In the extraction of iron from its ore which technique is used?

   a. Separating funnel   b. Centrifugation   c. Sublimation   d. All of these

36. Which is the following is a example of solids which sublime ?

   a. Ammonium chloride   b.Camphor   c. Antracene   d. All of these

37. How we can separate a mixture of salt and camphor ?

   a. Sublimation   b.Evaporation   c. Chromatograph   d. Filtration

38. Separation of dyes in black ink using ?

   a. Chromatography   b. Evaporation   c. Sublimation   d. Separating funnel

39. To separate drugs from blood which method is used?

   a. Centrifugation   b.Chromatography   c. Filtration   d. separating funnel

40. Which method is used for the separation of components of a mixture containing two miscible liquid?

   a. Distillation   b. Chromatography   c. Sublimation   d. None of these

41. How we can obtain pure copper sulphate from an impure sample?

   a. Chromatography   b. Crystallisation   c. Sublimation   d. Distillation

42. The method used in separation of crystals of alum from impure samples?

   a. Chromatography   b. Crystallisation   c. Sublimation   d. Distillation

43. Who was the first scientist to use the term element in 1661?

   a. Antoine Laurent   b. Robert boyle   c. Antoine Lavoiser   d. None of these

44. Which of the following is a properties of metal ?

a. Malleable   b. Ductile   c. Sonours   d. All of these

45. Which of the following a metal ?

a. Copper   b. Bromine   c. Iodine   d. Oxygen

46. Which is following a non metal?

a. Gold   b. Platinum   c. Silver   d. Carbon

47. Which is a metalloid?

a.Polonium   b. Germanium   c. Boron   d. All of these

48. Which two elements are liquid?

a. Mercury , lead   b. Mercury , Bromine   c. Bromine , Lead   d. Mercury, Oxygen

49. Which a compound out of a following?

a. Water   b. Copper   c. Stainless steel   d.Solder

50. Which of the following is physical change?

a. Burning of alcohol   b. Formation ice from water

c. Both (a) & (b)   d. None of these

Answers:

Q	A	Q	A	Q	A	Q	A	Q	A
1	B	11	A	21	C	31	A	41	B
2	C	12	B	22	B	32	B	42	B
3	A	13	A	23	A	33	A	43	B
4	C	14	D	24	C	34	D	44	D
5	D	15	B	25	B	35	A	45	A
6	D	16	A	26	D	36	D	46	D
7	A	17	B	27	A	37	A	47	D
8	D	18	A	28	D	38	A	48	B
9	C	19	C	29	C	39	B	49	A
10	B	20	D	30	D	40	A	50	B

3. ATOMS AND MOLECULES

SOME IMPORTANT POINTS

- Law of conservation of mass states that mass can neither be created nor be destroyed in a chemical reaction.
- In a pure substance the amount of elements present in a fixed proportion by mass. This is known as the law of constant proportion or the laws of definite proportion.
- The atomic theory was given by John Dalton. It tells that the atom is the smallest particle present in matter. Atoms are indivisible. Atoms of different elements are of different masses and different properties.
- A group of two or more atoms that are chemically bonded together are known as mo0lecules.
- The combining capacity of an atom is known as valency.
- The Avogadro constant represents the number of $6.022*10^{23}$.
- Mass (in gram) of 1 mol of a substance is called to molar mass.

3. ATOM AND MOLECULES

1. What is the meaning of atom?

   a. Derive   b. Indivisible   c. Divisible   d. Small

2. Who gave the laws of chemical combination?

   a. Dalton   b. L.Lavoisier   c. Joseph L. Proust   d. Both (a) & (b)

3. In water, the ratio of mass of hydrogen to the mass of oxygen is?

   a. 1:9   b. 14:3   c.1:8   d. 8:1

4. In ammonia the ratio of nitrogen and hydrogen in the ratio?

   a. 3:14   b. 14:3   c.13:4   d. 13:5

5. Who stated the law of constant proportion?

   a. L.Lavoisier   b. Joseph L.Proust   c. Dalton   d. Both (a) & (b)

6. Who give the basic theory about the nature of matter?

   a. E.Goldstien   b. J.J Thompson   c. John Dalton   d. Albert Einstien

7. Dalton theory based on?

   a. The law of conservation of mass.   b. The law of constant proportion

   c. The laws of chemical combination   d. The law of definite proportion

8. Atomic radius is measured in ?

   a. Picometer   b. Nano meter   c. Mano meter   d. None of these

9. How many nano meters in 1 meter?

   a. $1/10^9$ b. 10^{9nm} c. 10^{-9nm} d. None of these

10. Our entire world is made up of?

    a. Cell   b. Atom   c.Nucleus   d. All of these

11. The symbol of cobalt is?

    a. Cu   b. Co   c. Co   d. C

12. The symbol of lead is?

    a. PB   b. Pb   c. pb   d. Pa

13. The symbol of tungsten?

    a. Tu   b. Te   c. Cs   d. W

14. The symbol of Au is used for ?

    a. Silver   b. Arsenic   c. Gold   d. Argon

15. The atomic no. Of tungsten is?

    a. 53   b. 83   c.74   d. 89

16. The atomic no. Of iodine is?

    a. 49   b. 69   c.53   d. 43

17. Helium is a?

    a. Metaloid   b. Metal   c.Non metal   d. All of these

18. Earlier we defined the atomic mass unit as _____.

    a. u   b. v   c.amu   d. Am

19. In carbon monoxide, carbon and oxygen combined in the ratio of?

    a. 1:8   b. 3:4   c.4:3   d. 6:8

20. Who gave the atomic theory?

    a. E. Goldstien b. John Wolfgang c. John. Dalton d. Both (a) & (b)

21.   In these of the following which is very reactive?

        a. He   b. Xe   c. As   d. O

22.   How many isotopes of carbon exist in nature?

        a. 5   b. 2   c.12   d.15

23.   The atomic mass of the chlorine is?

        a. 35.9   b. 34   c.35.7   d. 35.5

24.   How many isotopes of hydrogen exist in?

        a. 0   b. 3   c.4   d. None of these

25.   The group of two or more atoms is known as?

        a. Element   b. Molecule   c. Compound   d. Mixture

26.   The atom of oxygen is known as ?

        a. Monatomic molecule   b. Tetra atomic molecule

        c. Diatomic molecule   d.Triatomic molecule

27.   The formula of ozone is?

        a. O_2 b. OZ   c.O_3 d. O

28.   The no. Of atoms consisting a molecule is known as its_____

        a. Monoatomic   b. Atom city   c.Molecity   d. Atomic mass

29.   Carbon is a?

        a. Metal   b. Non metal   c. Metaloid   d. None of these

30.   The ratio of masses of H:O is water is?

        a. 1:8   b. 1:2   c. 8:1   d. 3:8

31.   Carbon is a?

a. Cation   b. Anion   c. Both (a) & (b)   d. None of these

32.   What is the chemical formula of ammonium?

a.$HNCL_3$ b. $NHCL_4$ c. NH_3CL d. NH_4CL

33.   What is the formula of ammonium sulphate?

a. $(NH_4)SO_4$ b. NH_4SO_4 c. $(NH_4)_2SO_4$ d. $(NH)_4SO_{12}$

34.   What is the name of compound represent by formula $Al_2(SO_4)_3$?

a. Aluminium Sulphur   b.Argon sulphur

c. Aluminium Sulphate   d. AluminiumSulphide

35.   Formula of sodium nitrate?

a.NaNo   b. $NANO_3$ c.$NaNo_3$ d. $NaNo_3$

36.   What is the molecular mass of ammoia?

a. 16u   b. 13u   c. 8u   d. 18u

37.   What is the molecular mass of ammonium chloride?

a. 35.5u   b. 53.5u   c.55.5u   d.63.5u

38.   What is the molecular mass of potassium nitrate?

a. 100u   b. 99.2u   c.101.1u   d.101.3u

39.   The atomic mass of Argon is?

a. 29.98u   b. 39.9u   c.39.8u   d.39u

40.   Calculate the formula unit mass of K_2CO_3?

a. 128u   b.138.3u   c.138u   d.138.2u

41.   Molecular mass of CH_3OH?

a.31.22u   b.33.34u   c.32u   d.32.2

42.   Avogadro constant is representing?

a.N_1 b.AC   c.N_o d.AN_o

43.   What is the value of Avogadro constant is ?

a.$6.673*10^{-11}$ b. $6.667*10^{23}$ c. $6.023*10^{23}$ d. $6.022*10^{23}$

44.   How many dozen in 1 gross?

a. 1   b.13   c.6   d.12

45.   Gram atomic mass is known as?

a. Atomic mass   b. Atomicity   c. Molar mass   d. Mass number

46.   Calculate the no. of moles in 13g of hydrogen atom?

a.12   b.13   c.11   d.14

47.   Calculate the mass of 0.5 mole of N_2 gas?

a.7g   b.14g   c.12g   d.6g

48.   Calculate the no. of atom present in 2 moles?

a.$6.022*10^{23}$ b. $12.044*10^{23}$ c.$6.023*10^{23}$ d. $12.022*10^{23}$

49.   Calculate the no. of atoms present in 12g of hydrogen atom?

a. $72.022*10^{23}$ b.$72.264*10^{23}$ c.$6.022*12^{23}$ d.$72.04*10^{23}$

50.   Calculate the no. of atoms present in 69g of sodium atom?

a.$6.022*10^{23}$ b.$18.066*10^{22}$ c.$19*10^{23}$ d.None of these

Answers:

Q	A	Q	A	Q	A	Q	A	Q	A
1	B	11	C	21	D	31	B	41	C
2	D	12	B	22	B	32	D	42	C
3	C	13	D	23	D	33	C	43	D
4	B	14	C	24	B	34	C	44	D
5	B	15	C	25	B	35	C	45	C
6	C	16	C	26	C	36	D	46	B
7	C	17	C	27	C	37	B	47	B
8	B	18	C	28	B	38	C	48	B
9	B	19	C	29	B	39	B	49	B
10	B	20	C	30	B	40	D	50	D

4. STRUCTURE OF ATOM

SOME IMPORTANT POINTS

➢ The electron was identified by J.J Thomson and the proton was discovered by E. Goldstein.

➢ An atom has sub atomic particles, hence the atom is divisible.

➢ The charge on proton is equal in magnitude to the charge on electron but opposite in sign.

➢ The atomic nucleus was discovered by the Rutherford.

➢ The neutron was discovered by James Chadwick and it has no charge.

➢ The electron has negative charge, proton positive and neutron charge less.

➢ The mass of electron is about 1/2000 times of proton and the mass of proton and neutron is taken as 1 unit.

➢ The atomic number is the number of proton or electron in a neutral atom.

➢ Mass number is the number of proton and neutrons present in the nucleus of an atom.

➢ Isotopes are the atoms of same element which have same atomic number but different atomic mass.

➢ The first model for the structure of an atom was given by J.J Thomson.

4. STRUCTURE OF ATOM

1. The presence of new radiation in a gas discharge detected by J.J Thomson was?

   a. Anode   b. Cathode   c. Canal rays   d. All of above

2. Electrically charged object:

   a. Attract oppositely charged object   b. Attract uncharged object

   c. Repel uncharged object   d. None of these

3. Which one is not subatomic particle?

   a. Neutron   b. Neuron   c. Electron   d. All of above

4. Which one is negative charged atomic part?

   a. Proton   b. Neutron   c. Electron   d. All of above

5. The mass of a proton is taken as one unit and its charge as:

   a. Plus   b. minus   c. none of these

6. An electron is represented as:

   a. e^+ b. P^+ c. e^- d. P^-

7. The three main subatomic particles are?

   a. Neutron   b. Proton   c. Electron   d. All of these

8. The fundamental atomic particle having one unit positive charge is?

   a. Proton   b. Neutron   c. Electron   d. None of these

9. The ray's travels in straight line called?

   a. Cathode rays   b. X rays   c. Anode rays   d. None of these

10. _____ was the first one to propose model for the structure of an atom?

    a. Gaelili Galilio   b. J.J Thompson   c. Mark Henry   d. Albert Einstein

11. Cathode rays produce _____ when the strike against the surface of hard Metals?

a. X rays   b. Anode rays   c. Cathode rays   d. All of Above

12. Anode rays carry?

a. Positive charge   b. Negative charge   c. Neutral   d. None of these

13. In the model of J.J Thompson the rate of seeds in watermelon was as?

a. Proton   b. Electron   c. Neutron   d. None of these

14. An atom was a sphere of positive electrically in which were embedded numbers of electron sufficient to neutralize the positive charge?

a. Atom model of J.J Thompson   b. Atom model of Neils Bhor

c. Atom model of Rutherford   d. None of these

15. In the Thompsons model of atom, the sphere is positive charge is called ?

a. Proton   b. Electron   c. Neutron   d. All of above

16. The small heavy positively charged body present within the atom was called?

a. Neuron   b. Nucleus   c. Electron   d. Nucleus

17. Which type of particle Rutherford use in his experiment?

a. . α particle   b. B particle   c. γ particle   d. Atomic particle

18. During the experiment of Rutherford we used _____

a. Silver foil   b.Gold foil   c. Copper foil   d. Litmus paper

19. Rutherford's experiment α particle are doubly charged ?

a. Hydrogen ion   b. Helium ion   c. Argon ion   d. None of these

20. In the Rutherford α particle scattering gave totally unexpected result?

a. Most of α particle passed straight line through the gold foil

b. Some of α particle deflected in a small angle

c. Some of the α destroyed

d. Both (a) & (b)

21. According to Rutherford if your fire a 15 mg shells at a piece of tissue paper and it?

    a. Comes back and hit you   b. Go forward   c. Burn   d. None of these

22. The father of nuclear physics is?

    a. Stefen Howkins   b. E.Rutherford   c. Neils Bhor   d. Isaac Newton

23. According to Rutherford the size of nucleus is very small as compared to the size of?

    a. Nucleus   b. Atom   c. K Shell   d . α particle

24. Which order of shell is correct?

    a. M L K N   b. K L M N   c. L K M N   d. M N K L

25. The subatomic particle which revolves the nucleus?

    a. Protein   b. Neutron c. Electron   d. All of above

26. There is a gap in the atoms which is causes of?

    a. Passing of α particle   b. Deflection of α particle in same path

    c. α particle deflection in 180   d. All of above

27. The radius of the nucleus is about .... times less than the radius about the atom?

    a. 10^{11} b. 10^{5} c. 10^{12} d. 10^{9}

28. There is a positive charged centre in an atom called?

    a.Electron   b. Nucleus c. orbit   d. None of these

29. Which sub- atomic particle is Neutral?

a. Proton   b. Electron   c. Neutron   d. All of Above

30. Which of the level shell are first and last?

a. K&N   b. LM   c. LK   d. MN

31. The number of an element is equal to the number........ in the element ?

a. Electron   b. Proton   c. Neutron Proton   d.Neutron

32. Pb is symbol of?

a. Element   b. compound   c. mixture   d. None of these

33. Calculate the number of electron, Proton and Neutrons in sodium atom .

Given that atomic number of sodium is 11 and mass number is 23?

a. 20   b. 21   c.13   d. 12

34. Number of neutron is equal to?

a. Mass number – Atomic number   b. Atomic mass - Atomic number

c. Number of proton =Atomic number   d. None of these

35. Rutherford's scattering experiment showed for the first time that the atom has?

a. Electron   b. Proton   c. Nucleus   d. Neutrons

36. Proton was discovered by?

a. J.J Thompson   b. Rutherford   c. Chadwick   d. Goldstein

37. The atomic number of helium is?

a. 2   b. 1   c. 4   d. 13

38. The maximum number of shell present in the nth shell is equal to?

a. $n2^x$ b. $2n^2$ c. $2n^{x2}$ d. Xn^2

39. The electronic configuration of sodium is?

   a. K2 L8 M7   b. K2 L8 M1   c. K2 L8 M3   d. K2 L8 M2

40. The electron revolves the nucleus on its?

   a. Orbit   b. Triangle path   c. straight   d. None of these

41. The electron present in the outermost shell of the atom of an element is

   called?

   a. Valence shell   b. Valence electron c. Both (a) & (b)   d. None of these

42. The electronic configuration of chlorine is?

   a. K2 L8 M7   b. K2 L8 M5   c. K2 L8 M3   d. K2 L8 M6

43. The some element which have same atomic number but different mass number

    are called ?

   a. Isobar   b. Isotopes   c. Both (a) & (b)   d. None of these

44. The element which have some chemical property & different physical property?

   a. Isotopes   b. Isobar   c. Both (a)&(b)   d. None of these

45. The different elements which have different atomic number but same mass

   Number called?

   a. Isotopes   b. Isobar   c. Both (a)&(b)   d. None of these

46. Element with negative valency are?

    a. Always metals   b. Always metalloids

    c. Either metals or non metals   d. Always non metals

47. The first model of an atom was given by?

   a. N.Bohr   b. E.Goldstien   c. J.J Thompson   d. Rutherford

Answers:

Q	A	Q	A	Q	A	Q	A	Q	A
1	A	11	A	21	A	31	B	41	A
2	A	12	A	22	B	32	A	42	A
3	B	13	B	23	B	33	D	43	B
4	C	14	A	24	B	34	A	44	A
5	A	15	A	25	C	35	C	45	B
6	C	16	B	26	D	36	C	46	D
7	D	17	A	27	B	37	A	47	C
8	A	18	B	28	B	38	B		
9	C	19	B	29	C	39	B		
10	B	20	D	30	A	40	A		

5. CHEMICAL REACTION AND EQUATIONS

SOME IMPORTANT POINTS

➢ Chemical reaction has taken place if

1. Change in state

2. Change in colour

3. Evolution of gas

4. Change in temperature

➢ The symbolic representation of chemical reaction is chemical equation
➢ Mass can neither be created nor be destroyed in a chemical reaction
➢ The number of atom of each element is equal on both side of equation such equation called balanced chemical equation
➢ Those reaction in which two or more substance combine to form a new substances called combination reaction
➢ A compound splits in two or more substances such reaction called decomposition reaction
➢ Reaction in which heat is absorb called endothermic reaction
➢ Reaction in which heat is released called exothermic reaction
➢ Decomposition reaction carried out by :

1. Heat

2. Electricity

3. Sunlight

➢ Reaction in which one element takes place of another element in a compound are called displacement reaction
➢ Reaction in which two compounds react by the exchange of ions to form two new compound are called double displacement reaction
➢ Reaction in which addition of oxygen and loss of hydrogen in a substance oxidation reaction takes place

- Reaction in which addition of hydrogen and loss of oxygen in a substance reduction reaction takes place
- When a metal is attached by a substances it like moisture, air and water etc. And this process is called corrosion
- Nitrogen gas is flushed in packets of chips to prevent the chips getting oxidised

5. CHEMICAL REACTION AND EQUATIONS

1. Which of the following is Chemical reaction?

   (a) Milk is left a room temperature during summer

   (b) food digested in our body

   (c) Ice melts

   (d) both (a) and (b)

2. What would be formed when magnesium:

   (a) magnesium oxide   (b) Magnesium dioxide

   (c) magneseoxide   (d) none of these

3. Which of the following called skeletal Chemical equation

   (a) Balanced equation   (b) unbalanced equation

   (c) both (a) & (b)   (d) None of these

4. $Zn + H_2SO_4 \longrightarrow$ _____ + _____

   (a) $Zn_2SO_4 + H_2O$ (b) $ZnSO_4 + H_2$

   (c) $ZnSO_4 + H_2O$ (d) $Zn_2SO_4 + H_2$

5. $3Fe + 4H_2O \longrightarrow$ _____ + 4 _____

   (a) $Fe_3O_4 + H_2$ (b) $Fe_3O_4 + H_2O$

   (c) $Fe_2O_3 + H_2O$ (d) $Fe_2O_3 + H_2$

6. $CO + 2H_2 \xrightarrow{(340\ atm)}$ _____ + _____

   (a) $CH_3(OH)_3$ (b) CH_4 (c) CH_3OH (d) CH_2OH

7. $6CO + 6H_2O \xrightarrow{sunlight, chlorophyll}$ _____ + _____ $+6O_2$

   (a) $C_{12}H_{22}O_{11}$ (b) $C_6H_{12}O_6$ (c) $C_6H_{10}O_7$ (d) None

8. Fe + HCl $\longrightarrow$ _____ + _____

    (a) $FeCl_3 + H_2$ (b) $FeCl_2 + H_2$ (c) $Fe(Cl_3)_2 + H_2$ (d) $FeCl_2 + H_2O$

9. CaO + H_2O $\longrightarrow$ _____ + _____

    (a) CaOH   (b) $CaOH_2$ (c) Ca_2OH (d) $Ca(OH)_2$

10. What is the chemical formula of marble?

    (a) $Ca(OH)_2$ (b) $CaCO_2$ (c) $CaCO_3$ (d) $CaCO_4$

11. Which of the followings are used for the white washing of the walls?

    (a) Calcium Carbonate   (b) Calcium Hydroxide

    (c) Calcium Bicarbonate   (d) All of these

12. Which of the following is an example of exothermic reaction?

    (a) $CH_4 + 2O_2 \longrightarrow CO_2 + H_2O$

    (b) $2AgBr \xrightarrow{sunlight} 2Ag + Br_2$

    (c) $H_2O \xrightarrow{electricity} H_2 + O_2$

    (d) $2Ba(OH)_2 + NH_4Cl \longrightarrow 2BaCl_2 + NH_4OH$

13. which of the following reaction is used in black and white photography.

    (a) $Pb(NO_3)_2 \xrightarrow{heat} PbO + 4NO_2$

    (b) $2AgCl \xrightarrow{sunlight} 2Ag + Cl_2$

    (c) $AgBr \xrightarrow{sunlight} 2Ag + Br_2$

    (d) Both b and c

14. What is the colour of ferrous sulphat?

    (a) Blue   (b) Brown   (c) Green   (d) Yellow

15. which of the following is an example of displacement reaction?

(a) Fe + $CuSO_4$ $\longrightarrow$ $FeSO_4$ + Cu

(b) Zn + $CuSO_4$ $\longrightarrow$ CO_2 + H_2O

(c) Mg + $CuSO_4$ $\longrightarrow$ $MgSO_4$ + Cu

(d) All of these

16. The given equation is the example of:

$CaO + H_2O \longrightarrow Ca(OH)_2$ (aq.)

(a) Combination   (b) Decomposition

(c) Displacement   (d) Double Displacement

17. Which of the following is an example of double displacement reaction?

(a) $FeSO_4$ $\longrightarrow$ Fe_2O_3 + SO_2 + SO_3

(b) Pb + $CuCl_2$ $\longrightarrow$ $2PbCl_2$ + Cu

(c) Na_2SO_4 + $BaCl_2$ $\longrightarrow$ $BaSO_4$ + NaCl

(d) None of these

18. $2Pb(NO_3)_2$ $\xrightarrow{\text{heat}}$ $2PbO + 4NO_2 + O_2$

The above reaction is an example of?

(a) Combination   (b) Decomposition

(c) Displacement   (d) Double Displacement

19. Which type of reactions are called exothermic reactions?

(a) in which heat is released   (b) in which heat is absorbed

(c) Both a and b   (d) None of these

20. The decomposition reaction is carried out by applying:

(a) light   (b) heat   (c) electricity   (d) all of these

21. $Na_2SO_4 + BaCl_2 \longrightarrow BaSO_4 + 2NaCl$

In the above reaction, write the name of precipitate:

(a) Na_2SO_4 (b) $BaCl_2$ (c) $BaSO_4$ (d) NaCl

22. Which reaction is called oxidation reaction?

(a) in which oxygen is added   (b) in which hydrogen is removed

(c) Both a and b   (d) None of these

23. The reaction $2Cu + O_2 \xrightarrow{heat} 2CuO$

(a) Combination   (b) Oxidation   (c) Reduction   (d) Both a and b

24. What is the colour of the CuO in reaction?

$2Cu + O_2 \xrightarrow{heat} 2CuO$

(a) Blue   (b) Black   (c) Brown   (d) None of these

25. Which of the following is a example of redox reaction?

(a) $CuO + H_2 \xrightarrow{heat} Cu + H_2O$

(b) $ZnO + C \longrightarrow Zn + CO$

(c) Both a and b

(d) None of these

26. Which gives oxygen for oxidation or remove hydrogen is termed as ?

(a) Oxidizing agent   (b) Reducing agent

(c) Redoxing agent   (d) None of these

27. The black coating on silver is an example of:

(a) Rancidity   (b) Corrosion   (c) Reduction   (d) None of these

28. Which gas is used for protecting chips from rancidity?

(a) Argon   (b) Neon   (c) Hydrogen   (d) Nitrogen

29. Which reaction produces insoluble salts?

(a) Decomposition   (b) combination

(c) Precipitate   (d) Displacement

30. Which reaction occurs when an element displaces another element from its compound?

(a) Decomposition   (b) combination

(c) Precipitate   (d) Displacement

31. Which reaction is opposite of decomposition reaction?

(a) Combination   (b) Displacement   (c) Reduction   (d) Oxidation

32. Which of the following is an example of combination reaction?

(a) $Mg + O_2 \longrightarrow MgO$

(b) $NO + O_2 \longrightarrow NO_2$

(c) $H_2 + Cl_2 \longrightarrow HCl$

(d) All of these

33. Which of the following property a balanced chemical equation have?

a. Equal number of reactant and equal number of product.

b. Equal no. of atoms in reactants and products.

c. Different no. of atoms in reactants and products.

d. All of these

34. Which of the following is an oxidizing agent?

(a) $KMnO_4$ (b) $K_2Cr_2O_7$ (c) HNO_3 (d) All of these

35. $CuO + H_2 \longrightarrow Cu + H_2O$

In above reaction which substance is getting reduced?

(a) CuO   (b) Cu   (c) H_2O (d) H_2

36. Oxygen and moisture are necessary for?

(a) Corrosion   (b) Rusting of iron

(c) Both a and b   (d) None of these

37. A white coloured compound ……………..chloride changes grey when exposed to sun light.

(a) silver   (b) bromine   (c) barium   (d) none of these

38. $CaCO_3 \xrightarrow{\text{heat}} CaO + CO_2$

(a) Thermal decomposition reaction

(b) Electrolytic decomposition reaction

(c) Photolytic decomposition

(d) None of these

39. Complete the equation:

$Fe + HCl \longrightarrow$ _____ $+ H_2$

(a) $FeCl_2$ (b) $FeCl_3$ (c) Fe_2Cl_3 (d) Fe_3Cl_4

40. Burning of Coal is an example of?

(a) Exothermic reaction   (b) Endothermic reaction

(c) Decomposition   (d) none of these

41. Aluminium + sulphuric acid $\longrightarrow$ …………+ ……………

(a) Aluminium Sulphate, Water   (b) Aluminium Sulphate, Hydrogen gas

(c) Aluminium Oxide, Water   (d) Aluminium Hydroxide, Hydrogen gas

42. What is the colour of copper Sulphate?

(a) Blue   (b) Black

(c) Green   (d) Brown

43. In which reaction exchange of ions occurs?

(a) Combination reaction   (b) Decomposition

(c) Displacement   (d) Displacement

44. $H_2S + Cl_2 \longrightarrow$ ....+...................

(a) SO_2 +HCl   (b) $SCl_2 + H_2$

(c) S + HCl   (d) S_2 + HCl

45. $Fe + S \longrightarrow$ ....

(a) Fe_2S (b) FeS

(c) Fe_2S_3 (d) None of these

46. $Ca(OH)_2 + CO_2 \longrightarrow$ ....+ H_2O

(a) CaO   (b) $CaCO_2$

(c) $CaCO_3$ (d) $Ca(OH)_2$

47. What is the formula of slaked lime.

(a) CaO   (b) $Ca(OH)_2$

(c) $CaCO_3$ (d) None of these

48. $2Al + 3CuCl_2 \longrightarrow$ ....+ ....

(a) 2AlCl, 3Cu   (b) $2AlCl_3$, 3Cu

(c) $3AlCl_2$, $3Cu$ (d) None of these

49. $Zn + 2HCl \longrightarrow$ ....+ ....

     (a) $ZnCl_2$, H_2 (b) $ZnCl$, H_2

     (c) $ZnCl_2$, H_2O (d) $ZnCl$, H_2O

50. $3H_2 + N_2 \longrightarrow$ ....

     (a) $2NH_3$ (b) $2NH_4$

     (c) $2N_2H_3$ (d) $2NH_2$

ANSWERS:

QUE.	ANS.	QUE.	ANS.	QUE.	ANS.	QUE.	ANS.	QUE.	ANS.
1	D	11	B	21	D	31	A	41	B
2	A	12	A	22	C	32	D	42	A
3	B	13	D	23	D	33	B	43	D
4	B	14	C	24	B	34	D	44	C
5	A	15	D	25	C	35	A	45	B
6	C	16	A	26	A	36	C	46	C
7	B	17	C	27	B	37	A	47	B
8	A	18	B	28	D	38	A	48	B
9	D	19	B	29	C	39	B	49	A
10	C	20	B	30	D	40	A	50	A

6. ACIDS, BASES AND SALTS

SOME IMPORTANT POINTS

- Acids are sour in taste change blue litmus into red
- Bases are bitten in taste and change red litmus into blue
- Turmeric and red cabbage ate natural indicator whereas Methyl orange and phenolphthalein ate synthetic indicator
- Vanilla, onion and clove ate olfactory indicator
- Acid + metal $\rightarrow$ salt + hydrogen gas
- Acid + metal carbonate / metal hydrogen carbonate $\rightarrow$ salt + carbon dioxide + water
- Acid + base $\rightarrow$ salt + water
- This reaction is known as neutralization reaction like a base
- Non- metallic oxides are acidic in nature and behave like acid
- Acid and base release its ions water
- Acid release H^t, H_3O^t ions
- Base release OH^- ions
- Acid and base in water conduct electricity
- Bases which are soluble in water called alkalis
- The mixing of acid and base in water this reaction is highly exothermic
- The PH of a neutral solution is 7 and PH of acid is less than 7 and of base is more then 7.
- Our body works in PH 7.0 to 7.8 PH of rain water is less than 5.6 it is called acid rain tooth decay begins when PH of mouth is lower than 5.5
- Bleaching powder chemical name – calcium oxy chloride " formula – caocl2 used in textile paper Indus try and disinfecting drinking water
- Baking soda – chemical name – sodium hydrogen carbonate " formula $\rightarrow$ NAHCO3
- Washing soda – chemical name – sodium carbonate deca hydrates formula – $Na_2Co_3. 10H_2O$
- Water of crystallization is the fixed number of water molecules present in one formula unit of a salt

- ➢ Plaster of parts

  Chemical name – Calcium Sulphate hemihydrates

  Formula – $CaSO_4. \frac{1}{2}H_2O$

- ➢ Gypsum

  Chemical name – calcium sulphate dihydrate " formula – $CaSO_4. 2H_2O$.

6. Acid, Bases and Salts

1.   Acid is _____ in taste.

   a. Sweety   b. Bitter   c. Sour   d. None of these

2.   Acid changes the colour?

   a. Litmus Blue to Red litmus   b. Litmus Red to Blue litmus

   c. Litmus Red to Yellow   d. All of these

3.   A Strain of curry on a white shirt becomes reddish brown. when washed with deterg
   it becomes _____ in colour .

   a. Yellow   b. Red   c. Blue   d.Raddish

4.   Bases are _____ in taste?

   a. Sour   b. Bitter   c. Sweet   d. None of these

5.   Bases change the colour?

   a. Red litmus to Blue litmus   b. Blue litmus to Red litmus

   c. Radish Brown litmus to Red litmus   d. None of these

6.   Which are following a synthetic indicator for acids and bases ?

   a. Phenolphthalein   b.Lemon   c. Methyl orange   d. Both(a)&(b)

7.   When we react dilute sulphuric acid, added few pieces of zinc granules , then

   Pass the gas being evolved through the soap solutions them bubble will form

   , take burning candle near the bubble there would be pop sound so the gas

   Present the bubble is ?

   a. Oxygen   b. Hydrogen   c. Nitrogen   d. None of these

8.   The metals combine with the remaining part of the acid and form a compound

is called ?

a. Base   b. Acid   c. Salt   d. None of these

9.   The reaction of the acid with a metal can be summarised as?

a. Acid + Metal →Salt + Hydrogen gas

b. Acid + Salt → Metal + Hydrogen gas

c. Acid + Hydrogen gas → Metal + Salt

d. None of these

10.   Limestone, chalk and marble are different from of ?

a. Sulphuric acid   b. Calcium carbonate   c. Hydrocarbonate   d. None of these

11.   All metal carbonates and hydrogen carbonates react with _____ to give a

Corresponding salt, carbon dioxide and water

a. Acid   b. Salt   c. Bases   d. None of these

12.   The reaction between an acid and a base to give a salt and water is known as?

a. Neutralisation   b. Oxidation   c. Addition   d. All of these

13.   The general reaction between a metal oxide and an acid can be written as?

a. Base + Acid → Salt + Water   b. Metal oxide + Acid → Salt + Water

c. Acid + Salt → Water + Metal   d. All of these

14.   What happen when small amount of copper oxide in a beaker and add dilute

Hydrochloride acid ?

a. The solution becomes blue – green   b. The metallic oxide change in gas

c. The copper oxide dissolve .   d. Both (a)&(c)

15.   The reaction of a base with an acids metallic oxides are said to be?

a. Phynolic oxide   b. Acidic oxide   c.Basic oxide   d.All of these

16.   When we touch the blue litmus the colour will change into?

a. Yellow   b. Red   c. Green   d. Reddish blue

17.   Acid found in?

a. Lemon   b. Curd   c. Oranges   d. All of these

18.   Which gas is usually liberated when an acid reacts with a metal?

a. Hydrogen gas   b. Oxygen gas   c. Nitrogen gas   d. Helium gas

19.   When we mix acid with base element remains?

a. Salt   b. Base   c. Acid   d. Acid remains   d. All of these

20.   Choose wrong statement?

a. All acids have similar chemical properties

b. All acids have different chemical properties

c. Acids are sour in taste   d. Base change the colour of items red litmus

21.   Which element is common for all acids?

a. Oxygen   b. Hydrogen   c. Helium   d. Argon

22.   Which of solution do not conduct electricity ?

a. Glucose   b. Salty water   c. Alcohol   d. Both (a)&(c)

23.   The electric current is carried through the solution by?

a. Ions   b. Molecules   c. Isotopes   d. All of these

24.   The cation present in acids is ?

a. H^+ b. H^- c.E^- d. E^+

25.   Acids produce hydrogen ions, H^+ (aq), in solution which have are responsible

For?

a. Basic property   b. Acidic property   c. Alcoholic property   d.None of these

26.   Which of the chemical is alkali?

a. Barium hydrochloride   b. Sodium hydrochloride

c. Both (b)&(d)   d. Calcium hydrochloride

27.   Choose the chemical formula of hydrochloric ion is?

a. H_4O^+ b. H_2O c. H^+O_4 d. H_3O

28.   Hydrogen ions cannot exist after alone, but they exist after combing with?

a. Air molecule   b. Hydrogen molecule

c. Water molecule   d. None of these

29.   Hydrogen ions cannot exist alone , but they exist after combing with water

molecules , Thus hydrogen ions must always be shown as H^+(aq) or hydrogen

ion (H_3O^+) is summarise as ?

a. $H^+ + 2HO \rightarrow H_3O^+$ b. $H^+ + H_2O \rightarrow H_2O$

c. $H^+ + H_2O \rightarrow H_3O^+$ d. None of these

30.   Acids gives _____ or _____ in water.

a. H_3O^+ or H^+(aq)   b. H_2O or H^+(aq)   c. H_3O or H_2O d. None of these

31.   Bases generate _____ in water?

a. Hydrogen   b. Oxygen   c. Hydrochloride   d. None of these

32.   Which are soluble in water are called?

a. Alkalis   b. Acidic   c. Alkali   d. All of these

33.   The processing of dissolving an acid or a base in water is highly?

a. Exothermic   b. Intrinsic   c. Acidic   d. Basic

34.   Human stomach produces?

a. Hydrochloric acid   b. Hydrochlorine   c. Bases   d. All of these

35.   To get rid of stomach pain, people uses bases called ?

a. Hydrochloric acid   b. Sulphuric acid   c. Antacids   d. All of these

36.   The ph scale measure concentration of _____ in a solution ?

a. Oxygen   b. Nitrogen   c. Helium   d. Hydrogen

37.   The ph scale of a neutral is?

a. 6   b. 6.9   c. 7   d. 7.2

38.   Value less than 7 on the ph scale represents on ?

a. Basic solution   b. hydrochloric solution

c. Acidic solution   d. None of these

39.   The ph values less than 7 on the ph scale representation an increase in _____ in

ion?

a. OH⁻   b. E⁻   c. OH⁺   d. All of these

40.   In the ph scale 7 to 14 represents _____ nature

a. Acidic   b. Basic   c. Hydrochloric   d. None of these

41.   If we take hydrochloric acid of the same concentration, sexy one molar, then

these produce _____ amount of hydrogen ions

a. Same   b. Different   c. Big   d. None of these

42.   Acid that gives rise to more H⁺ ions are said to be _____

a. Strong acids   b. Weak acids   c. Acid in solid   d. All of these

43. When ph of rain is less than _____ is called acid.

   a. 6   b. 7   c. 5.6   d. 8.2

44. Human`s tooth made up of hardest substance in the body is _____

   a. Calcium   b. Calcium   c. Calcium phosphate   d. None of these

45. Stinging hair of nettle leaves inject _____ cause burning pain.

   a. Methanoic acid   b. Ethanoic acid   c. Base   d. None of these

46. Acid present in curd is?

   a. Ethanoic acid   b. Citric acid   c. Acetic acid   d. Lactic acid

47. Which of the following is the formula of salt?

   a. Nacl   b. Cl H_2O c. Na_2SO_4 d. Both(a) &(c)

48. Salt of a strong acid and a strong base are with ph value of?

   a. 3   b. 7   c. 8   d. 5.5

49. The common salt thus obtained is an important raw material for various materials of daily use, such as ?

   a. Baking soda   b. Sodium hydrochloride

   c.Bleaching powder   d. All of these

50. When electricity is passed solution of sodium hydrochloride. This process called ?

   a. Chlor- Alkali   b. Chlor – Alkali   c. Aldehyde   d.None of these

51. Bleaching powder is represent as ?

   a. 2NaoH(aq)   b. $H_2(g)$ c. 2NaCl(aq)   d. $CaoCl_2$

52. Choose the correct use bleaching powder ?

a. For bleaching cotton and linen   b. As an oxidising agent in chemical industry

c. Potable water to make free of germ   d. All of above

53.   On heating gypsum at 373 k it losses water molecule and becomes calcium

Sulphate hemihydrates ($CaSO_4$ ½ H_2O)this is called ?

a. Base   b. washing   c.Baking soda   d. Plaster of Paris

Answers :

Q	A	Q	A	Q	A	Q	A	Q	A	Q	A
1	C	10	B	19	A	28	C	37	C	46	D
2	A	11	A	20	B	29	C	38	C	47	C
3	A	12	A	21	B	30	A	39	A	48	B
4	B	13	B	22	D	31	C	40	B	49	D
5	A	14	D	23	A	32	A	41	B	50	A
6	D	15	C	24	A	33	B	42	A	51	D
7	B	16	B	25	B	34	A	43	C	52	D
8	C	17	D	26	C	35	C	44	B	53	D
9	A	18	A	27	D	36	D	45	A		

7. METAL AND NON METAL

SOME IMPORTANT POINTS

- Metals are lustrous, malleable, ductile, sonorous, hard, goods conductors where as non-metals are not
- Non-metals are brittle, generally low melting point
- Iodine is non-metal but it is lustrous
- Diamond is non-metal but it is hardest natural substance in the would
- Lead and mercury are metals bat poor conductors of heat and electricity
- Graphite is non-metal but it is conduct electricity
- Lithium, sodium, potassium can be cut by a knife
- metals have low ionization energy it is the energy is the amount of energy required to remove the electrons from the valance shell
- metal + oxides $\rightarrow$ metal oxide
- metal oxides react with acid as well as base are called amphoteric oxides
- aqua regia is freshly prepared mixture of concentrated hydrochloric acid and concentrated nitric acid in the ration of 3:1
- ionic compound are hard, soluble in water, rigid structure and conduct electricity in molten state
- the ore obtained after mining from the ground contain large amount of can they and sandy impurities called gangue
- cinnabar is ore of mercury
- calcination it is process of heating carbonate ore in lack of air
- roasting it is process of heating sulphide ore in excess of air
- thermit reaction used for welding of railway tracks
- galvianisation is a method of protecting steal and iron from rusting by coating then with a thin layer of zinc
- an alloy homogeneous mixture of two or more metal or a metal and metal
- solder is alloy of lead and tin has a low melting point used for welding electrical wires.

7. METAL AND NON METAL

1. Metals have shining surface. This property is called?

    a.Sonours   b.Hardness   c. Malleable   d. Metallic lusture

2. Metals can be beaten into thin sheets .This property is called _____?

    a.Sonours   b. Hardness   c.Malleable   d.Metallic lusture

3. Which of the following metals is most malleable metals?

    a.Gold   b.Silver   c. Lead   d.Both (a)&(b)

4. The ability of metals to be drawn into thin wire is called _____?

    a.Sonours   b.Ductility   c.Malleability   d.Regidity

5. 1 gm of gold can be drawn to from a wire of about _____ length.

    a.2000km   b.2000m   c.2000dm   d.None of these

6. Which metals are comparatively poor conductors of heat?

    a.Lead and Mercury   b.Mercury and silvers

    c.Mercury and copper   d.Lead and copper

7. The Wires that carry in your homes which material is coated on them?

    a.Polychloride   b.Polyvinychloride   c.Polyviny   d.None of these

8. Which of the following non metal is good conductor of electricity?

    a. Carbon   b.Iodine   c. Graphite   d.Diamond

9. Which of the following metal is a liquid at room temperature ?

    a. Mercury   b.Bromine   c.Xenon   d.None of these

10. Which metals have very low melting point gallium?

    a.Mercury   b.Caesium   c.Arsenic   d.Both(a)&(b)

11.   _____ is non metal but it is lustrous

a.Carbon   b.Iodine   c.Graphite   d.Radon

12.   Which metal have low density?

a.Lithium   b. Boron   c.Silicon   d.Aluminium

13.   Which oxides are produced when non metal are dissolved in water?

a.Basic   b.Acidic   c.Both(a)&(b)   d.None of these

14. $2Cu+O_2 \rightarrow$ _____

a.CuO   b.2CuO   c.$2Cu_2O$ d.Cu_2O

15.   Which metals oxides which react with both acids as well as bases to produce

Salts and water?

a.Aluminium oxide   b.Zinc oxide   c.Both (a)&(b)   d.None of these

16. $Al_2O_3+2NaOH \rightarrow$ _____$+H_2O$

a. $2NaAlO_2$ b.$2NaAlO_3$ c.$2NaAlO$ d.$2Na_3Al_2O$

17.   Which of the following is correct chemical equation?

a.$Na_2O+H_2O \rightarrow NaH+O_2$ b.$Na_2O+H_2O \rightarrow NaOH$

c.$K_2O+H_2O \rightarrow K(OH)_2+H_2$ d.$K_2O+H_2O \rightarrow K(OH)_2$

18.   Which metal catch fire when kept in open?

a.K,Na   b.Na,I   c. Na,P   d. K,I

19.   Which Process is forming a thick layer of aluminium ?

a.Rusting   b. Anoding   c.Both (a)&(b)   d.Galvanisation

20. $Al_2O_3+6 HCL \rightarrow$ _____ + _____

a.$2AlCl_3 + 4H_2O$ b.$3AlCl + 3H_2O$ c.$2AlCl_3 + 3H_2O$ d. None of these

21.   Which of the following react with cold water ?

   a.Calcium   b. Magnesium   c.Zinc   d. Aluminium

22. $3Fe + 4H_2O \rightarrow$_____ $+ 4H_2$

   a.Fe_2O_3 b.Fe_3O_4 c.Fe_4O_5 d.Fe O

23.   Which of the following metal do not react with water at all ?

   a. Iron   b. Aluminium   c. Sodium   d. Lead

24.   Which of the following is more reactive ?

   a. Iron   b. Aluminium   c. Sodium   d. Lead

25.   Aquar regia is a freshly prepared mixture of ?

   a.Dilute Hydrochloric acid and Dilute Nitric Acid

   b.Concentrated Hydrochloric Acid and Concentrated Nitric Acid

   c.Concentrated Hydrochloric Acid and Dilute Nitric Acid

   d.Dilute Hydrochloric Acid and Concentrated Nitric Acid

26.   Which of the following in correct for reactivity?

   a.Al>Zn>Pb>Cu   b.Zn>Al>Pb>Cu   c.Al>Zn>Cu>Pb   d.Zn>Al>Cu>Pb

27.   Which element has electronic configuration 2,8,8,1?

   a. K-   b. Ar   c. K+   d.Ca+

28.   Which of the following is a ionic compound?

   a.N_2 b.NaCl   c.$MgCl_2$ d.Both (b)&(c)

29.   Ionic compound soluble in ?

   a.Kerosene   b.Petrol   c.Water   d.None of these

30.   Which of the following element most aboudan on earth crust is _____.

a.Copper   b.Aluminium   c.Silver   d.Gold

31.   Which of the following element found in free state ?

       a.Sodium   b.lead   c.Copper   d. Zinc

32.   The ore obtained after mining from the ground contains a large amount of

       Earth and Sandy impurities .What they called?

       O_2e b.Gangue   c.Mineral   d.None of these

33.   Cinna bar is an ore of _____ .

       a.Mercury   b.Lead   c.Silver   d.Iron

34. $ZnCo_3$ connecting into oxide which process is used ?

       a.Roasting   b.Calcination   c.Anodising   d.None of these

35. $3MnO_2 + 4Al \rightarrow 3$_____ $+2$_____

       a.$Al_2O_3 + Mn$ b.$AlO_3 + Mn_2$ c.$Mn + Al_2O_3$ d.$Mn_2 + AlO_3$

36.   Which of the following reaction in thermit reaction?

       a.$3MnO_2 + Hal$ --------- _____ $3Mn + 2Al_2O_3$

       b.$Fe_2O_3 + 2Al$ ---------- $2Fe + Al_2O_3$

       c.$2Hgs + 3 O_2$ ---------- $2Hgo + 2So_2$

       d.None of these

37.   The metal obtained is not pure is called _____

       a.Curd Metal   b.Crude Metal   c.Curdy Metal   d. Surde Metal

38.   Whichof the following method used in ?

       a. Galvanising   b.Annodising   c.Both(a)&(b)   d.None of these

39.   The alloy of measuring is _____ .

a. Cinnabar   b.Amalgam   c.Solder   d.None of these

40.   Which Metal oxide is used in preventing rusting of iron for iron pillar near

Qutub minar, New Delhi?

a. Magnetic oxide   b.ZnO   c.Fe_2O_3 d.Al_2O_3

41.   Steel is alloy of?

a. Iron , Carbon   b.Iron ,Silicon   c.Iron ,Nickel   d.None of these

42.   Which alloy is used in welding of wires?

a. Amalgam   b.Solder   c.Steel   d.Cinnabar

43.   Solder is an alloy of?

a. Lead ,copper   b.Copper,Tin   c.Lead,Tin   d.Tin,Copper

44.   Which of the following alloy in rust free ?

a. Stainless Steel   b.Steel   c.Cinnabar   d.Amalgam

45.   Bronze is an alloy of?

a.Copper and Zinc   b.Copper and Tin c.Copper   d.Copper and Mercury

46.   Iron,Nickel and Chromium are consistuent of which alloy ?

a.Stainless steel   b.Steel   c.Cinnabar d.None of these

47.   Which colour coating on copper carbonate when reacts with moist CO_2 ?

a.Blue   b.Brown   c.Green   d.Black

48.   In thermit reaction which element powder is used as reducing agent?

a.Mn   b.Fe   c.Zn   d.Al

49. $Mg + 2HNO_3 \rightarrow$ _____ + _____

a.$MgNo_3 + H_2$ b.$MgNO_3 + H_2$ c.$MgCNO_3 + H_2$ d.$Mg(NO_3) + H_2O$

50. $ZnO + C \rightarrow$ _____ + _____

   a. $Zn + Co_2$ b. $Zn + Co$ c. $Zn + Co_3$ d. None of these

Answers :

Q	A	Q	A	Q	A	Q	A	Q	A
1	D	11	B	21	A	31	A	41	A
2	C	12	A	22	B	32	B	42	B
3	D	13	B	23	D	33	A	43	C
4	B	14	B	24	C	34	B	44	A
5	B	15	C	25	B	35	C	45	B
6	A	16	A	26	A	36	B	46	A
7	B	17	B	27	D	37	B	47	C
8	C	18	A	28	D	38	C	48	D
9	A	19	B	29	C	39	B	49	C
10	D	20	C	30	B	40	A	50	B

8. CARBON AND ITS COMPOUND

SOME IMPORTANT POINTS

➤ The earth's crest has only 0.02% carbon in the form of mineral
➤ Most of carbon compounds are poor conductor of electricity and have low melting and boiling point
➤ The force of attraction between these molecules are not very strong
➤ Bonding in compounds are mainly covalent bond
➤ The two property of carbon tetra valency and catenation gives rise to large number of molecule
➤ The sharing of electron is known as covalent bond
➤ Carbon compound are two types saturated and unsaturated
➤ Saturated compounds has single bonds between the carbon atoms and unsaturated was double and triple bonds between carbon atoms
➤ Carbon compounds having identical molecular formula but different structure known as structural isomers
➤ The saturated hydrocarbons also called alkanes
➤ The carbon having double bond called alkene and triple bond called alkynes
➤ The formula of

$$\text{Alkanes} : C_N H_{2N+2}$$
$$\text{Alkenes} : C_N H_{2N}$$
$$\text{Alkynes} : C_N H_{2N-2}$$

➤ In hydrocarbons the element replaced the hydrogen is reffered as heteroatom
➤ These at heteroatom confer specific properties to the compound are called functional group
➤ A homologous series is a group of organic compounds having similar structure and similar chemical properties and successive compound is differ by CH_2 group
➤ When carbon compounds burnt in oxygen to give carbon dioxide along with the release of heat light

- Saturated hydrocarbons burns with a clean flame while unsaturated carbon compounds will give a yellow flam with lots of black smoke
- carbon compounds can be easily oxidized alcohols are converted to carboxylic acids by oxidation
- the industrial use of addition reaction in making vegetable
- the boiling of ethanol is 35k it is colourless liquid with pleasant small and it is miscible in water
- combustion of ethanol gives carbon dioxide and water + heat
- ethanol reacts with sodium to form sodium oxide and H_2
- the reaction between Ethanol and ethanoic acid is known as es trification and its reverse reaction is known as saponification
- the common name of ethanoic acid is acetic acid
- the dilute solution of acetic acid (5-8) in water is called vinegar
- the melting point of acetic acid is 190k and boiling is 391k
- ethanoic acid reacts with base to form salt and water
- it reacts with metal carbonates and hydrogen carbonates to form salt water and CO_2
- it is used in the form vinegar and as a food preservative for pickles sausages etc
- soaps are sodium or potassium salts of long chain of carboxylic acid
- detergents ate ammonium or sulphonate salts of long chain of carboxylic acid
- detergent is used in both hard and soft water but soap is used in only soft water and soap is bio-degradable

8. CARBON AND ITS COMPOUNDS

1. The earth's crust has carbon in the form of minerals of about:

   a) 5%   b) 3%   c) 0.02%   d) 0.03%

2. The ionic compounds have melting and boiling points at?

   a) High rate   b) low rate   c) Neutral   d) none of these

3. The melting point of Methane is:-

   a) 40k   b) 60k   c) 90k   d) 80k

4. The combining capacity of various elements dependss on the number of:-

   a) Protons   b) Valence electrons   c) Isotopes   d) None of these

5. The atomic number of carbon is:-

   a) 5   b) 4   c) 6   d) None of these

6. The reactivity of elements is explained as their tendency to attain a:-

   a) Noble gas Configuration   b) Ionic compound c)   Carbonic   Compound
   d) None of these

7. Carbon attains the noble gas configuration by

   a) Replacing electrons   b) Sharing electrons

   c)Graining electrons   d) None of these

8. How many electrons, helium has in its K shell

   a) 5 electrons   b) 4 electrons   c) 2 electrons
   d) None of these

9. How many electrons required by oxygen to complete its octet:-

   a) 2 electrons   b) 3electrons

   c) 8 electrons   d) none of these

10. The bond formed between two oxygen atoms is a:-

a) Single bond   b) double bond   c) Triple Bond
d) None of these

11.   A major Component of biogas is:-

a) Ethane   b) Pentane   c) Methane   d) None of these

12.   The word "Tetravalent" means:-

a) 4electrons in valence shell   b) 4 electrons in K Shell
c) 8 electrons in Valence shell   d) None of these

13.   The Covalent bonds are such bonds which are formed by:-

a) Sharing of electrons between two atoms

b) Isotopes

c) None of these

14.   Covalently bonded molecules have:-

a) High melting points   b) Low melting points

c) Neutral   d) None of these

15.   The atomic number of nitrogen is 7, its valency is:

a) 4   b) 5   c) 3   d) None of these

16.   Compound of carbon which is linked by single bonds is called:-

a) Saturated Compound   b) Unsaturated Compound
c) Flactuated Compound   d) None of these

17.   Compound of Carbon which is linked by double bonds is called:-

a) Saturated Compound   b) Unsaturated Compound
c) Flactuated Compound   d) None of these

18.   Compounds of carbon which are linked by triple bonds are called:-

a) Saturated Compound   b) Unsaturated Compound
c) Flactuated Compound   d) None of these

19.   The versatile nature of carbon consist

a) Tetravalency   b) Catenation   c) Both (a) & (b)   d) None of these

20.   Friedrich Wohler Prepare Urea by which compound:-

a) Ammonium Sulphate   b)AmmoniumCyanide
c) Sodium Carbonate   d) None of these

21.   The Saturated Compounds are normally:-

a) Highly Reactivity   b) Less reactive   c) Neutral   d) None of these

22.   The Structural Isomers Consists:-

a) Straight Carbon Chain   b) Branched Chain

c) Cyclic chains   d) All of these

23.   A Carbon Compound Methene is the member of

a) Alkane group   b) Alkene group   c) Alkyne Group   d) None of these

24.   A Compound of carbon, Haxane has the bond of:-

a) Single bond   b) Double bond   c) Triple bond   d) None of these

25.   The hydro Carbon compounds consists:-

a) Carbon   b) Hydrogen c) Both (a) & (b)   d) None of these

26.   The atom of element which replace the hydrogen atoms is called:-

a) Isotope   b) Heteroatom   c) Neutrons   d) None of these

27.   The first member of Alkane group is:-

a) Methene b) Ethane   c) Ethene   d) None of these

28.   The first member of Ketone group is:-

a) Methene b) Propanone   c) Butanone   d) None of these

29.   The first member of carboxyllicalid is:-

a) Methanal b) Ethanoicacid   c) Methanoic acid d) None of these

30.   The isomer of Aldehyde group is:-

a) Ethanol   b) Ethanal   c) Propanone   d) None of these

31.   Which is remain similar in a hhomologous series:-

a) Chemical Property   b) Melting Point   c) Static Store   d) None of these

32.   The Chemical formula of Alkene is:-

a) C_nH_{2n+2} b) C_nH_{2n} c) C_nH_{2n-2} d) None of these

33.   The Chemical formula of Alkane is:-

a) C_nH_{2n+2} b) C_nH_{2n} c) C_nH_{2n-2} d) None of these

34.   The functional group of chloro-propane is:-

a) Aldehyde b) Ketone   c) Halogen   d) Carboxyllicacid

35.   Which suffix is present in members of ketone:-

a) -yne   b) -one   c) -oicacid   d) -al

36.   When Carbon Compound, burn in air it gives:-

a) Carbon dioxide b) Heat   c) Light   d) All of these

37.   Which hydrocarbons when burn gives a yellow flame with lots of black smoke:-

a) Saturated hydrocarbons   b) Unsaturated Hydrocarbon

c) Complex Compounds   d) None of these

38.   Which hydrocarbons give sooty flame when burning:-

a) Saturated hydrocarbons   b) Unsaturated Hydrocarbon

c) Complex Compounds   d) None of these

39.   Which Catalyst is using in oxidizing reaction:-

a) Potassium   b) Alkaline Potassium Permanganate   c) Nickel d) None of these

40.   Vegetable oils generally have long chains of:-

a) Saturated Carbon   b) Unsaturated Carbon

c) Complex Carbon   d) None of these

41. Which catalyst is using in Addition reaction:-

a) Nickel   b) Ammonia   c) Copperd) None of these

42. What is added to iodine to make it tincture iodine:-

a) Carboxyllicacid   b) Ethanol   c) Ammonium   d) None of these

43. Which is a physical property of Ethanol:-

a) Melting Point   b) Boiling Point   c) Good Solvent   d) All of these

44. Which plants are the convertors of sunlight into chemical energy:-

a) Wheat Plant   b) Cotton plant

c) Sugarcane Plant   d) None of these

45. When ethanol reacts with sodium it forms:-

a) Sodium Ethoxide   b) Sodium Ethanoate   c) Sodium hydroxide
      d) All of these

46. When ethanoic acid reacts with sodium it forms:-

a) Sodium ethoxide   b) Sodium ethanoate

c) sodium hydroxide   d) All of these

47. Which is as Physical property of Ethanol acid:-

a) Sour tasteb) Melting Point   c) Boiling Point   d) All of these

48. Ethanoic acid reacts with ethanol to give:

a) Potassium   b) Ester   c) Sodium Saltd) None of these

49. Soaps are long chain carboxylic acid of:-

a) Ammonium Salts   b) Nickel Salts   c) Potassium Salts   d) None of these

50. The molecule of soaps consists:-

a) Hydrophilic ends

b) Hydrophobic ends

c) Both (a) & (b)

d) Sulphonate ends

Answers:

1. (c)	2. (a)	3. (c)	4. (b)	5. (c)	6. (a)	7. (b)	8. (c)	9. (a)	10. (b)
11. (c)	12. (a)	13. (a)	14. (b)	15. (c)	16. (a)	17. (b)	18. (b)	19. (c)	20. (b)
21. (b)	22. (d)	23. (b)	24. (a)	25. (c)	26. (b)	27. (c)	28. (b)	29.(c)	30. (b)
31. (a)	32. (b)	33. (a)	34. (c)	35 (b)	36. (d)	37. (b)	38. (a)	39. (b)	40. (b)
41. (a)	42. (b)	43. (d)	44. (c)	45. (a)	46. (b)	47. (d)	48. (b)	49. (c)	50. (c)

9. PERIODIC CLASSIFICATION OF ELEMENTS

SOME IMPORTANT POINTS

➤ Elements are classified on their similarities and differences.

➤ Dobereiner was the first scientist to show the relationship between the properties and their atomic masses.

➤ Law of triads states that when elements are arranged in order of increasing atomic masses in a group of three elements, the atomic, mass of middle element of the triads being equal to arithmetic mean of the atomic mass of the two elements.

➤ Newland's law of octaves states that when elements are arranged in order of increasing atomic masses, the properties of every eighth element is similar to the first element.

➤ Mendeleev arranged the element in order of increasing atomic mass and their chemical properties.

➤ Mendeleev left some spaces for few undiscovered elements in his periodic table.

➤ The anomalies of Mendeleev's periodic table are removed in Modern Periodic Table.

➤ In Modern Periodic Table going left to right in periods the atomic size and metallic properties decreases.

➤ In Modern periodic table going up to dawn in group, valency remains same, metallic properties and size of atom increases.

9. PERIODIC CLASSIFICATION OF ELEMENTS

1. In which year Doberenier made the triads

   a) 1917   b) 1817   c) 1816   d) 1815

2. Which of the following is not a traids

   a) Li, Na, K   b) Ca, Sr, Ba   c) Cl, Br, I   d) All of these

3. How many elements were found when Newland Classify elements:-

   a) 56   b) 55   c) 50   d) 57

4. Law of octane's applicable up to -----------

   a) Thorium   b) Calcium   c) Magnesium   d) Tungsten

5. How many elements were found in Mendeleev's time?

   a) 56   b) 63   c) 73   d) 53

6. Write the oxide of the tungsten.

   a) WO_2 b) WO_3 c) W_2O_3 d) W_7O_7

7. The Outer shells of the elements are completely filled are known as:-

   a) Nobel gas   b) Inert gas   c) Metalloid   d) both (a) & (b)

8. The atomic Mass of Silicon is :-

   a) 28   b) 28.1   c) 28.2   d) 28.3

9. The atomic No. Start from 58 to 71 is known as:-

   a) actinides b) lanthanides   c) both (a) & (b)   d) none of these

10. The vertical columns are known as:

    a) Group   b) Rows   c) Period   d) both (a) and (b)

11. The elements belongs to the 17 group are known as:-

a) alkenes   b) actinides   c) Halogens d) both (a) & (b)

12. The atomic number of the elements is 14 is belong to the group in the periodic table.

a) 12   b) 13   c) 14   d) 9

13. The atomic radius is measured into

a) meter   b) Pm   c) um   d) hm

14. Atomic radius of Hydrogen atom is:

a) 37pm   b) 30pm   c) 37m   d) 30m

15. 1Pm = --------m

a) 10^{12}m   b) 10^{-12} c) 10^{-11} d) 10^{-10}

16. Which of the following is a non-metal

a) Iodine (I)   b) Hg   c) SC   d) Cs

17. Which of the following is a metalloid?

a) Re   b) Ir   c) Sb   d) Se

18. Modern periodic table is based on -------------

a) Atomic Mass   b) Increasing atomic number

c) Increasing atomic mass   d) Both (a) & (c)

19. Which of the followings is Reactive?

a) Rn   b) As   c) Xe   d) None of these

20. According to Mendeleev's periodic law, elements are arranged in

a) the increasing order of atomic mass

b) decreasing order of atomic mass

c) increasing order of atomic number

d) decreasing order of atomic number

21. Which of the following statement is correct about mendelliv's periodic law:-

    a) It has 18 Horizontal rows known as periods

    b) It has 6 horizontal rows know as periods

    c) It has 18 vertical columns know as group

    d) It has 7 vertical rows know as group

22. Which of the given elements A,B,C,D, and E with the atomic number 11,37,16,18, and 5 belong to the some group

    a) A,B,C   b) A,B   c) D,E,C   d) A,C,E

23. Where you would place element in the periodic table which electronic Configuration is 2,8,8:-

    a) 17th group   b) 18th group   c) 2nd group   d) 13 group

24. An element which is an essential constituent of all organic compounds belongs to

    a) 1st group and 2nd period   b) 14th group, 2nd period

    c) 14th group, 1st period   d) 13th group, 3rd period

25. Which of the following is the outer most shell for element of group 3?

    a) K Shell   b) M shell   c) N Shell   d) All of these

26. Which of the following elements exhibit maximum no. of valency electron.

    a) Cr   b) w   c) Zn   d) both (a) & (b)

27. Which of the following element exhibit maximum no. of Shells?

    a) Cr   b) Ne   c) Zn   d) Both (a) & (c)

28. Which of the following element has the largest number of radius?

    a) Po   b) As   c) O   d) W

29. Which among the following has the more metallic character?

    a) As   b) Be   c) Ba   d) both a and c

30. Among the following elements in the order of their decreasing atomic size Ag, Au, V, Ge, Ba is?

a) Ag>Ag>V>Ge>Be   b) Ge<V<Ag<Au<Ba   c)   Ba>Au>Ag>V>Ge
d) both (a) and (c)

31. What type of oxide would EKa-Boron forms?

a) BO_3 b) B_3O_2 c) B_2O_3 d) BO

32. These elements B, Si and Ge are:-

a) Metals   b) Non-Metals
c) Semi-Metal   d) Non-Metal, Metalloid and metals respectively

33. Which of the following element from an acidic oxide?

a) An element belongs to 15th group and 6th period

b) An element belongs to 14th group and 3rd period

c) An element belongs to 4th group and 4th period.   d) None of these

34. Which one of the following doesn't increase while moving down the group of the periodic table?

a) atomic radius   b) Metallic character   c) Valency   d) No. of shells

35. Write the Hydride forms by Eka-Sillico

a) SH_2 b) SH_4 c) S_2H_3 d) S_2H_5

36. The Atomic number gives us ------------------------

a) The Mass of Atom   b) No. of Protons present in the nucleus

c) No. of electrons   d) both (b) and (c)

37. How many noble gases discussed in the modern periodic table.

a) 7   b) 6   c) 5   d) 4

38. The electronic configuration of an element is 2,8,8,6

a) 5th period and 6th group   b) 4th period and 6th group

c) 4th period and 5th group   d) 3rd period and 3rd group

39. Which of the oxide forms by tungsten?

a) TgO_3 b) WO_3 c) W_2O_7 d) Tg_2O_7

40. The atomic radius of oxygen is:

a) 74pm   b) 63Pm   c) 66Pm   d) 72Pm

41. Which of these the atomic radius is largest

a) C   b) Li   c) Be   d) N

42. the atomic size refers to

a) Atomic Number   b) Atomic Mass

c) Atomic radius   d)atomic structure

43. Debierne's Classify only ----------- elements

a) 3   b) 6   c) 9   d) 12

44. Which of this element has no neutron is its nuclear:

a) Cl   b) As   c) H   d) He

45. The electronic configuration of H⁻ is

a) 2,1   b) 1   c) 2   d) 2,8,8,8,3

46. Which of this act like as metal as well as non metal?

a) Na   b) Cl   c) W   d) H

47. The atomic No. of Barium is

a) 37   b) 35   c) 56   d) 45

Answers:

QUE.	ANS.	QUE.	ANS.	QUE.	ANS.	QUE.	ANS.	QUE.	ANS.
1	B	11	C	21	B	31	C	41	B
2	D	12	C	22	B	32	C	42	C
3	A	13	B	23	B	33	B	43	C
4	B	14	A	24	B	34	C	44	C
5	B	15	B	25	D	35	D	45	C
6	B	16	A	26	C	36	D	46	D
7	D	17	C	27	D	37	B	47	C
8	B	18	B	28	D	38	B	48	
9	B	19	B	29	C	39	B	49	
10	A	20	A	30	B	40	C	50	

NOTES

www.ingramcontent.com/pod-product-compliance
Lightning Source LLC
Chambersburg PA
CBHW080600180526
45168CB00007B/2725

* 9 7 8 1 5 1 6 8 2 0 9 7 9 *